PENGUIN READERS

Dear Parents and Educators,

Welcome to Penguin Readers! As parents and educators, you know that each child develops at their own pace—in terms of speech, critical thinking, and, of course, reading. As a result, each Penguin Readers book is assigned an easy-to-read level (1–4), detailed below. Penguin Readers features esteemed authors and illustrators, stories about favorite characters, fascinating nonfiction, and more!

EMERGENT READER
Simple vocabulary • Word repetition • Picture clues • Predictable story and sentence structures • Familiar themes and ideas

PROGRESSING READER
Longer sentences • Simple dialogue • Picture and context clues • More in-depth plot development • Nonfiction and fiction

TRANSITIONAL READER
Multisyllable and compound words • More dialogue • Different points of view • More complex storylines and characters • Greater range of genres

FLUENT READER
More advanced vocabulary • Detailed and descriptive text • Complex sentence structure • In-depth plot and character development • Full range of genres

*This book has been officially leveled by using the F&P Text Level Gradient™ leveling system.

For Dashka and Jojo,
my anchors in any storm
—SJ

PENGUIN YOUNG READERS
An imprint of Penguin Random House LLC
1745 Broadway, New York, New York 10019

First published in the United States of America by Penguin Young Readers,
an imprint of Penguin Random House LLC, 2025

TIME for Kids © 2025 TIME USA, LLC. All Rights Reserved.

Photo credits: cover, 3: PeopleImages/iStock/Getty Images; 4: Khanh Bui/Moment/Getty Images; 5: (left, top) StockPlanets/E+/Getty Images, (left, bottom) Lyn Walkerden Photography/Moment/Getty Images, (right) Abstract Aerial Art/DigitalVision/Getty Images; 6: Brian Bonham/500px Prime/Getty Images; 7: Cheunghyo/Moment/Getty Images; 8: miljko/E+/Getty Images; 9: Oleh_Slobodeniuk/E+/Getty Images; 10: (top) Peter Cade/Stone/Getty Images, (bottom) Sunphol Sorakul/Moment/Getty Images; 11: Bruce Wilson Photography/iStock/Getty Images; 12: Antoninapotapenko/iStock/Getty Images; 13: Eloi_Omella/E+/Getty Images; 14: KonArt/iStock/Getty Images; 15: Slavica/E+/Getty Images; 17: Enrique Díaz/7cero/Moment/Getty Images; 18: janiecbros/E+/Getty Images; 19: mdesigner125/iStock/Getty Images; 20: Novarc Images/Jonas Piontek/mauritius images GmbH/Alamy Stock Photo; 21: (top) PeopleImages/E+/Getty Images, (bottom) puhimec/iStock/Getty Images; 22: akrassel/iStock/Getty Images; 23: (left) Colors Hunter/Chasseur de Couleurs/Moment/Getty Images, (right) JillianCain/iStock/Getty Images; 24: cokada/E+/Getty Images; 25: Hannah Bichay/DigitalVision/Getty Images; 26: (left) Eerik/iStock/Getty Images, (right) msan10/iStock/Getty Images; 27: Jenny Dettrick/Moment/Getty Images; 28: ConstantinCornel/iStock/Getty Images; 29: Danny Lehman/The Image Bank/Getty Images; 30: Jason E. Vines/Moment/Getty Images; 31: Byrdyak/iStock/Getty Images; 32: amriphoto/E+/Getty Images; 33: Jure Batagelj/500px/500px Prime/Getty Images; 34: Chad Cowan/500px Prime/Getty Images; 35: Tokarsky/iStock/Getty Images; 36: antonyspencer/E+/Getty Images; 37: Science Photo Library/NOAA/Brand X Pictures/Getty Images; 38: john finney photography/Moment/Getty Images; 39: mack2happy/E+/Getty Images; 40: chuchart duangdaw/Moment/Getty Images; 41: David Bailey/Nashville/DigitalVision/Getty Images; 42: JJ Gouin/iStock/Getty Images; 43: Emad aljumah/Moment/Getty Images; 44: sshepard/E+/Getty Images; 45: Tatiana Gerus/Moment/Getty Images; 46: M Swiet Productions/Moment/Getty Images; 47: Phillip Espinasse/500px/500Px Plus/Getty Images; 48: Photos by R A Kearton/Moment/Getty Images

Penguin supports copyright. Copyright fuels creativity, encourages diverse voices, promotes free speech, and creates a vibrant culture. Thank you for buying an authorized edition of this book and for complying with copyright laws by not reproducing, scanning, or distributing any part of it in any form without permission. You are supporting writers and allowing Penguin to continue to publish books for every reader.

Visit us online at penguinrandomhouse.com.

Library of Congress Cataloging-in-Publication Data is available.

Manufactured in China

ISBN 9780593752722 (pbk)   10 9 8 7 6 5 4 3 2 1 WKT
ISBN 9780593888032 (hc)   10 9 8 7 6 5 4 3 2 1 WKT

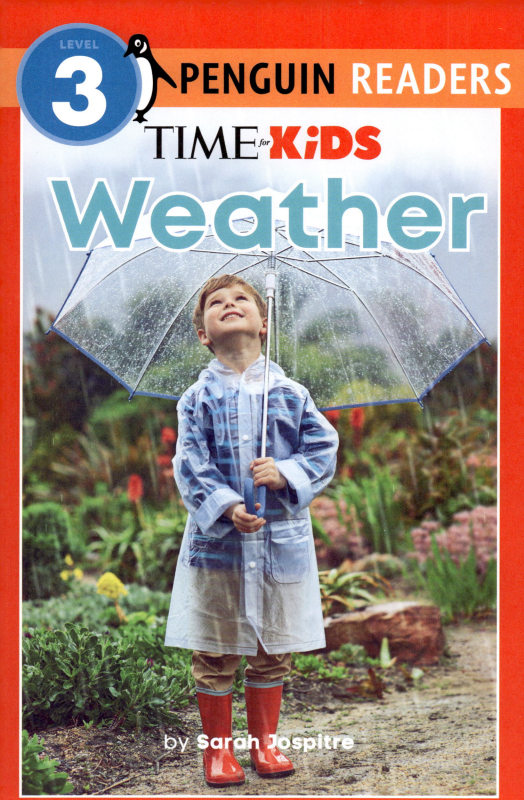

Ghostly clouds of fog spreading through a town.

The perfect snow day.

Massive rain puddles ready to be jumped in.

Whirling dust devils.

These are all results of weather. But what *is* weather?

Weather refers to the different conditions that happen in the atmosphere. Weather can be dry, wet, cold, hot, stormy, calm, clear, or cloudy. Weather is different all around the world.

Let's take a look at some of the most remarkable weather conditions!

Fog

Have you ever wanted to walk through a cloud? Well, chances are you have—through fog, that is!

Similar to clouds high in the sky, fog is a cloud that touches the ground. It is made up of water droplets or ice crystals that form water vapor, which is water in the form of gas. When water vapor turns back into liquid, or condenses, the air cools. Bits of dust in the air, very

humid and hot weather, or cold and moist weather, and gentle winds allow fog to appear. Sometimes, these clouds are so thick that they can be hard to see through.

Fog can also form when cool air meets warm, moist air over water or land; when warm air blows over a cool body of water; or when rain falls from warm air to cool air near the ground.

Fog can be seen in valleys and coastal areas near bodies of water. It usually forms at night or in the early morning when the air is cool. In colder places, fog can last the entire day.

Wind

Wind is the movement of air. From light breezes to hurricanes, wind is caused by changes in the temperature of water, land, and air. Some of the most powerful winds whip up as cyclones and tornadoes.

Winds in Antarctica can reach speeds of 200 miles per hour. That's faster than some sports cars!

Weather overall is greatly affected by the wind. Wind brings warm air into cool areas, and can even carry rain, snow, dust, and sand into places that did not have them before. For example, winds such as monsoons in South Asia bring rain during the summer. This is because warm, moist air from the ocean moves in toward cooler land.

For as long as there have been people, we have used wind to help us in many ways. For example, in 5000 BCE, wind power was used to move boats along the Nile River, in Africa. Today, windmills are used to power machines that create electricity.

Lightning and Thunder

Lightning is a very big spark of electricity in the sky. When it strikes the ground, lightning is very dangerous to anyone and anything nearby. Because lightning usually hits the tallest objects, people should stay far away from trees and mountains during storms.

Did you know that about 100 lightning bolts strike Earth every second? In the United States, there are about 25 million lightning strikes per year.

Ice particles and water droplets inside a cloud carry positive and negative electrical charges. The droplets with positive charges float to the top of the cloud while those with negative charges sink to the bottom. When too much electrical charge builds up and separates in a cloud, it creates an electrical field where lightning occurs. When the positive and negative charges combine, a bright streak known as a lightning flash is created. This flash can be as hot as 54,000°F. That is about five times hotter than the sun's surface! The process of the negative and positive charges building up and then combining repeats until there are not enough charged particles to produce lightning.

Do you know what the odds of being struck by lightning are? One in 15,300!

Depending on what the light travels through, lightning can even be seen in different colors. For example, during snowstorms, lightning can appear pink and green.

Thunder is the loud sound caused by lightning. It is so loud because of the intense heat that lightning generates. When electricity from lightning heats the gases in the air, the gases grow bigger and make a loud, thunderous sound. The louder the sound of thunder, the closer you are to lightning.

Since light travels faster than sound, we see lightning before we hear the thunder. It can sound like anything from a low rumble to a powerful *boom*!

The place that gets the most lightning on Earth is believed to be Lake Maracaibo, in Venezuela. The lake has lightning nearly half the days of the year.

Rain

Rain is water in liquid form that falls from the sky. When clouds become saturated, or filled with water droplets, raindrops fall from them. Rain is the main source of fresh water in the world, and can be found almost everywhere on Earth.

One of the world's rainiest places is Mount Waialeale, in Hawaii. There, it rains about 350 days a year!

All living things need water to survive. Too little or too much rainfall can be very bad. When there is not enough rainfall, droughts can leave organisms thirsty and likely to die. Too much rainfall can lead to flooding, and can drown plants and animals.

One of the driest places on Earth is the Atacama Desert, in Chile. It gets almost no rainfall. Some parts of this desert have not had rain in hundreds of years!

Ice: Hail, Snow, and Blizzards

Ice is the solid, or hard, form of water. It is found in weather conditions such as hail, snow, and blizzards.

Picture tons of ice cubes falling from the sky. This is how hail falls to the ground. Hailstones are pieces of ice that fall out of the sky. Any type of water that falls from clouds is called precipitation. Hail, rain, snow, and sleet are all examples of precipitation.

Hailstones form when cold rain gets lifted back up into the coldest parts of a storm and frozen in the atmosphere before falling back to the ground. It may be hard to see very far when hail is falling.

Tiny crystals of ice that fall to the ground are called snow. When these ice crystals clump  together, they form snowflakes. Snowflakes form all sorts of unique shapes and can look like a star, a needle, or a plate, depending on the temperature at the time they form. One snowflake can contain more than 100 ice crystals!

During the winter, snow falls in places that get cold weather. It helps the environment and living organisms. Because snow reflects most of the sun's heat, snow on the ground helps keep the planet cool. It also provides water for animals and people when it melts in the spring. But if snow does not melt for many years, glaciers, or huge sheets of ice, form.

When cold temperatures, powerful winds, and plenty of snow combine, we get strong snowstorms called blizzards.

In the United States, they mostly occur between December and February, and can last for more than three hours. Roads and transportation systems may shut down, and power lines may stop working. In cold places such as Antarctica where powerful winds blow, blizzards are frequent. When a blizzard makes it difficult to see where the ground ends and the air begins, it is called a whiteout.

Storms: Thunderstorms, Tornadoes, and Cyclones

Storms are powerful disruptions in the atmosphere that bring wind and sometimes rain, snow, lightning, and thunder.

One of the most common storms is a thunderstorm: a storm with thunder and  lightning that often also includes thick clouds, heavy rain or hail, and strong winds.

Thunderstorms can be violent but typically do not last long. They occur when moist, hot air quickly rises to cooler parts of the atmosphere, and they happen in every country in the world. As the air cools and clouds and rain droplets form, lightning develops inside the clouds. Lightning is the most dangerous feature of a thunderstorm. Air expands as lightning quickly heats it, causing the sound of thunder. Very quickly, air cooled by the rain rushes toward the ground and causes heavy winds.

Certain thunderstorms are powerful enough to create  tornadoes, and even a series of tornadoes called a tornado family. Tornadoes are storms where powerful swirling

winds form a column from the clouds to the ground. These storms are the smallest but most violent disruptions on Earth, with winds as fast as 300 miles per hour.

Although they can occur almost anywhere on Earth, most reported tornadoes each year happen in the Great Plains region of the United States. Several tornadoes that form over a broad region are called a tornado outbreak or tornado family. During the Super Outbreak of 2011, more than 300 tornadoes formed in 15 states across America.

A tropical cyclone, such as a hurricane, is a strong spinning storm of powerful winds and rain that begins over warm ocean waters near the equator. Some cyclones stay out over  the sea, while others pass over land. They weaken when they spend too much time on land or over cool waters.

Tropical cyclones are dangerous because they cause flooding and can even pick up big things in their way, such as boats! The center of a cyclone is called the eye of the storm. A hurricane can unleash over 2.4 trillion gallons of rain per day!

Heat: Heat Waves, Droughts, and Dust Devils

Extreme heat is a time of high temperatures over 90°F. When above-normal temperatures, with or without high humidity (high levels of water vapor in the atmosphere), last for more than two days, this weather event is called a heat wave. A heat wave can damage crops, injure and kill livestock, increase the risk of wildfires

(an uncontrollable fire in a forest, grassland, or other ecosystem), and result in power shortages and blackouts. Heat waves can also decrease the water supply.

Another weather event that can affect the water supply is a drought. Droughts occur when there is a shortage of rain over time. They can be harmful to people, plants, and animals. They occur due to changes in the Earth's atmosphere, such as shifts in the wind and changing ocean currents.

Droughts can occur anywhere in the world, at any time, and can be extreme or normal. For example, places that experience rainy and dry seasons usually have a seasonal drought during the drier months. The worst droughts can last for months or years. This makes it hard for crops to grow and leads to plants, animals, and even people dying of famine, or lack of food.

A dust devil is a short but strong whirlwind that occurs when the surface of the Earth is heated by the sun. As the ground heats, it warms the air above it. This warm air rises, then begins to spin, creating the whirlwind motion of a dust devil.

Dust devils can be seen because of the swirling dust, leaves, and other loose debris they pick up. Though they are small, dust devils can damage or destroy small structures like houses. They occur in places like Arizona, where there is little rain and droughts are more likely.

Rainbows

A rainbow is a multicolored band in the sky. It forms when light from the sun shines through water. This leads to light reflecting and refracting (or bending) off water droplets and splitting into seven colors: red, orange, yellow, green, blue, indigo, and violet. This series of colors is called a light spectrum.

Rainbows are full circles, not arcs. From the ground, we only see *half* of the circle.

Depending on such factors as the angle of the sun and the size of the water droplets in the air, a rainbow can last from a few minutes to an hour. Rainbows can form anywhere in the world, but they are most common in tropical areas, such as the island of Kauai, in Hawaii. Hawaii is known as the Rainbow State.

Earth is the only planet in the solar system with rainbows. That's because Earth is the only planet where liquid precipitation and direct sunlight are consistent.

Weather comes in many different shapes, sizes, forms, and degrees of danger—from mild and sweet, like the rainbow after a thunderstorm, to dangerous and even disastrous. What's the weather like where you are?